Cambridge Primary

Hodder Cambridge Primary
Maths

Activity Book
B

Foundation Stage

Ann and Paul Broadbent

HODDER
EDUCATION
AN HACHETTE UK COMPANY

Every effort has been made to trace all copyright holders, but if any have been inadvertently overlooked, the Publishers will be pleased to make the necessary arrangements at the first opportunity.

Although every effort has been made to ensure that website addresses are correct at time of going to press, Hodder Education cannot be held responsible for the content of any website mentioned in this book. It is sometimes possible to find a relocated web page by typing in the address of the home page for a website in the URL window of your browser.

Hachette UK's policy is to use papers that are natural, renewable and recyclable products and made from wood grown in sustainable forests. The logging and manufacturing processes are expected to conform to the environmental regulations of the country of origin.

Orders: please contact Bookpoint Ltd, 130 Park Drive, Milton Park, Abingdon, Oxon OX14 4SE. Telephone: (44) 01235 827720. Fax: (44) 01235 400401. Email: education@bookpoint.co.uk Lines are open from 9 a.m. to 5 p.m., Monday to Saturday, with a 24-hour message answering service. You can also order through our website: www.hoddereducation.com

© Ann Broadbent and Paul Broadbent 2018

First published in 2018

This edition published in 2018 by Hodder Education,
An Hachette UK Company
Carmelite House
50 Victoria Embankment
London EC4Y 0DZ
www.hoddereducation.co.uk

Impression number   10 9 8 7 6 5 4 3 2 1

Year   2022 2021 2020 2019 2018

Cover illustration by Steve Evans

Illustrations by Jeanne du Plessis, Vian Oelofsen

Typeset in FS Albert 17 pt by Lizette Watkiss

Printed in the United Kingdom

A catalogue record for this title is available from the British Library.

ISBN 978 1 5104 3183 6

Contents

Numbers to 10

Count the dots. Say the number.

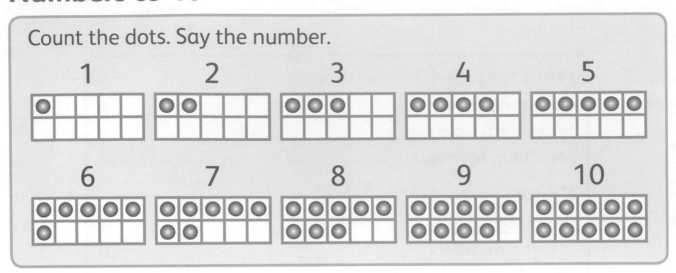

⭐ Read the number on each door. Count the flowers. Write the numbers.

Writing numbers

⭐ Trace over these numbers. Start at the dots and then write the numbers.

1	1	1	1			
2	2	2	2			
3	3	3	3			
4	4	4	4			
5	5	5	5			
6	6	6	6			
7	7	7	7			
8	8	8	8			
9	9	9	9			
10	10	10	10			

Matching numbers

⭐ Count the bees in each group.
Join each group of bees to the matching number on a hive.

⭐ Draw some flowers in these pots.
Write the number of flowers on each pot.

Counting

 Play this counting game.

How to play:
- Make a spinner using a paper clip and pencil as shown on page 9.
- Each player puts a counter on START.
- Take turns to spin the paper clip on the spinner and move your counter up the mountain.
- Stop at the top when you reach 10.
- On your next turn, count from 10 to 1 backwards as you move the counter down the track.
- The first player to reach HOME is the winner.
- Continue the game until all players have returned to HOME safely.

WELL DONE!

Stop at the top

Make a spinner like this.

Lost a hat, go back 2

Fell over, go back 1

HOME

Missing numbers

Here is a number track to 10.

| 1 | 2 | 3 | 4 | 5 | 6 | 7 | 8 | 9 | 10 |

⭐ Fill in the missing numbers on each track.

1	2			5		7		9

2	3	4		6		8	9	

	2	3	4				8	

2			5	6	7			10

⭐ Read what the train driver says below.
Colour the numbers on the track.

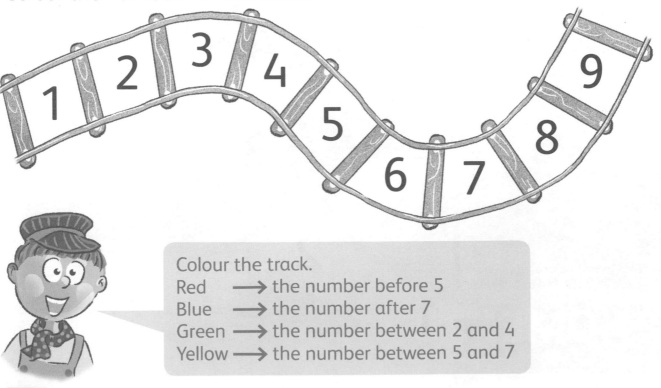

Colour the track.
Red ⟶ the number before 5
Blue ⟶ the number after 7
Green ⟶ the number between 2 and 4
Yellow ⟶ the number between 5 and 7

Counting on

The rabbit starts at 3 and counts on 4.
Where does it finish?

⭐ Show the jumps and write the finish number each time.

Start at 5 and count on 3 ⟶ ☐

Start at 2 and count on 5 ⟶ ☐

Start at 3 and count on 6 ⟶ ☐

Start at 6 and count on 4 ⟶ ☐

Finding totals

Two numbers are added together to make a total.

1	2	3	4	5	6	7	8	9	10

3 and 4 make 7

⭐ Find these totals.

1	2	3	4	5	6	7	8	9	10

[ ] and [ ] make [ ]

1	2	3	4	5	6	7	8	9	10

[ ] and [ ] make [ ]

1	2	3	4	5	6	7	8	9	10

[ ] and [ ] make [ ]

1	2	3	4	5	6	7	8	9	10

[ ] and [ ] make [ ]

Here are two ways to make 6.

3   and   3   4   and   2

Can you find another way?

⭐ Write numbers to make 7.

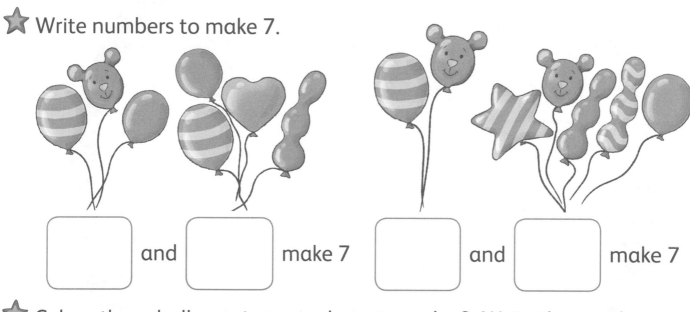

[ ] and [ ] make 7   [ ] and [ ] make 7

⭐ Colour these balloons in two colours to make 9. Write the numbers.

[ ] and [ ] make 9

Partitioning

⭐ Use 8 counters.

Put some counters on each side of this plate to make 8.

⭐ Write some of the ways you made 8.

8 is ☐ and ☐   8 is ☐ and ☐

Count the cubes.

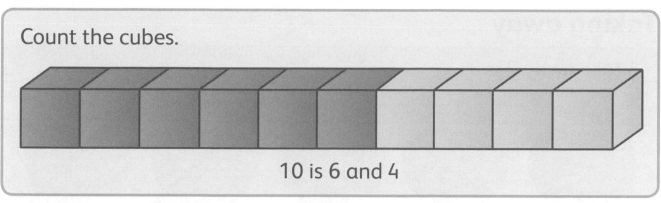

10 is 6 and 4

⭐ Use cubes in 2 colours. Make different ten-sticks. Colour these cubes to match your sticks. Write how you made 10 for each stick.

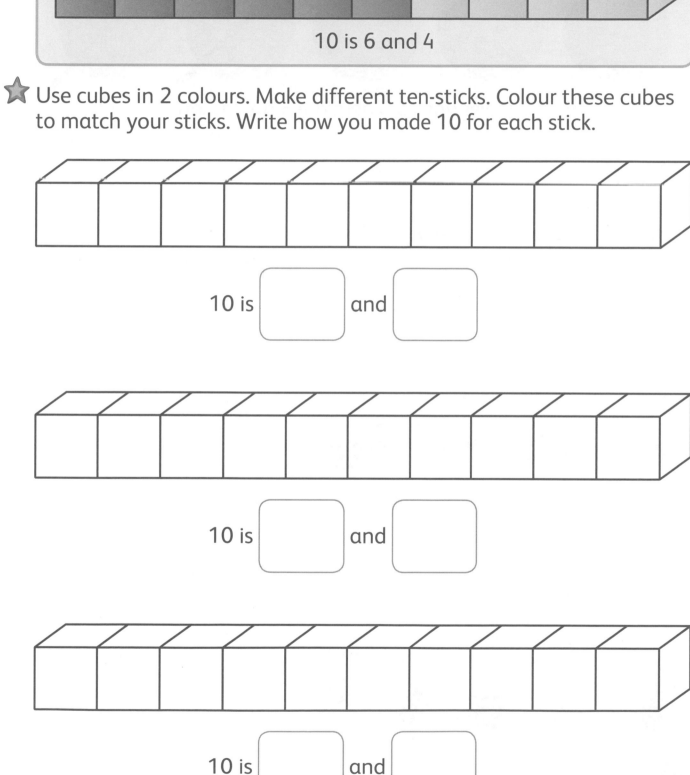

10 is ☐ and ☐

10 is ☐ and ☐

10 is ☐ and ☐

Taking away

5 take away 2 is 3

There are now 2 fewer apples. There are 3 apples left.

⭐ Cross out 2 fruit in each row. Write how many are left.

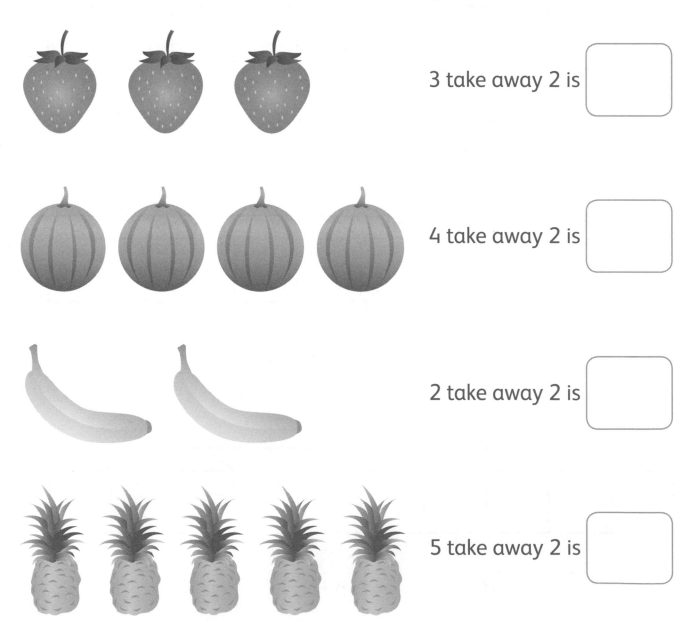

3 take away 2 is ☐

4 take away 2 is ☐

2 take away 2 is ☐

5 take away 2 is ☐

⭐ Put 5 counters on the snake.
Take counters away and write the number.

5 take away 1 is ⬚

5 take away 2 is ⬚

5 take away 3 is ⬚

5 take away 4 is ⬚

5 take away 5 is ⬚

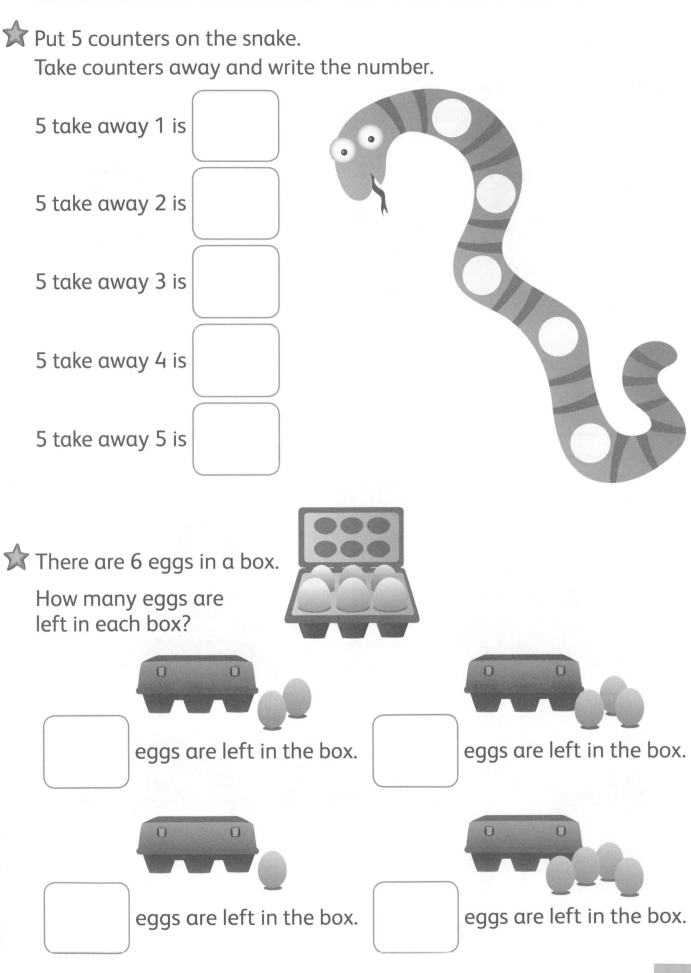

⭐ There are 6 eggs in a box.

How many eggs are
left in each box?

⬚ eggs are left in the box.

⬚ eggs are left in the box.

⬚ eggs are left in the box.

⬚ eggs are left in the box.

Story sequences

⭐ Talk about these pictures.
Write 1, 2 and 3 on the pictures in each story to show the order.

Months

 Which of these months do you know?
Choose a special event to draw in the middle of the circle.
Join your event to its matching month.

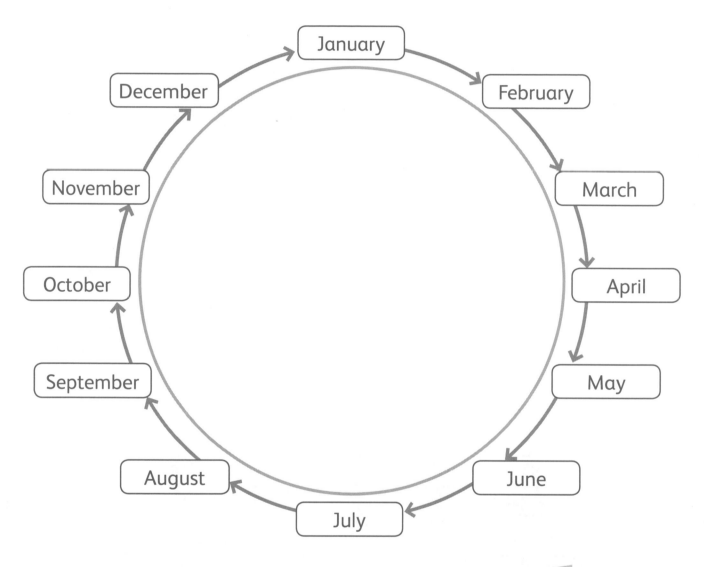

 Write the month of your birthday.

The month of my birthday is

_____.

Position of objects

⭐ Look at this toy shop.
Which toy is **below** the ball?
Which toy is **next to** the duck?

⭐ Which toy is **between** the drum and the boat?
Which toy is **above** the books?

Opposites

⭐ Join the opposites.

in

open

short

empty

closed

out

full

long

⭐ Draw opposites.

narrow   wide   down   up

short   tall   little   big

off   on

Length

 Tick the longest in each group.

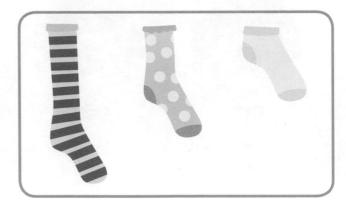

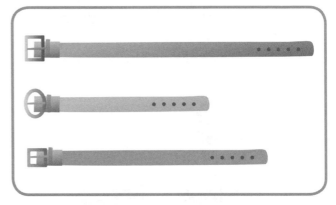

 Find things shorter than this pencil.
Draw them.

Ordering objects

⭐ Use cubes to make these cube sticks.
Join these cube sticks in order of length. Start with the shortest.

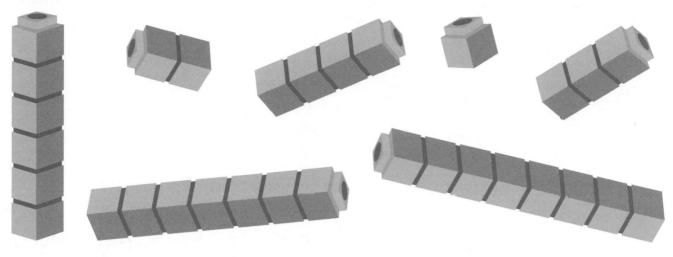

⭐ Draw lines to put these parcels in order of size. Start with the smallest.

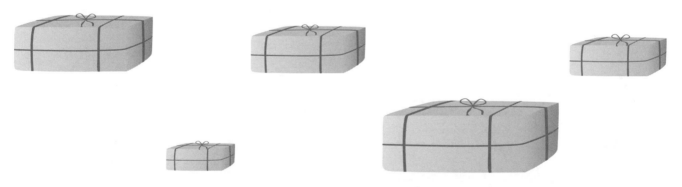

⭐ Number these jugs in order from 1 to 4. Number 1 is the most full.

Pictures and patterns

Colours, shapes and lines can make different patterns.

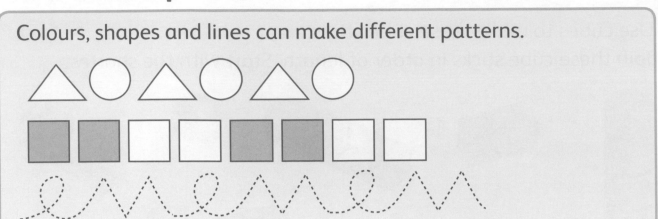

⭐ Trace over these patterns. Colour the shapes to make a pattern.

⭐ Draw 4 more shapes in each row. Colour the shapes to make patterns.

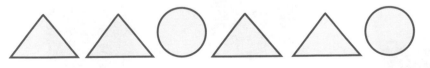

Flat shapes

Try to learn the names of these shapes.
Count the number of sides on each shape.

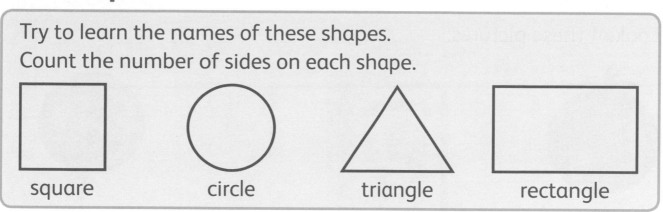

square   circle   triangle   rectangle

⭐ Draw a line to join each object to a matching shape.

Shapes around us

⭐ Look at these pictures.

How many are shaped like boxes?

How many are round and shaped like balls?

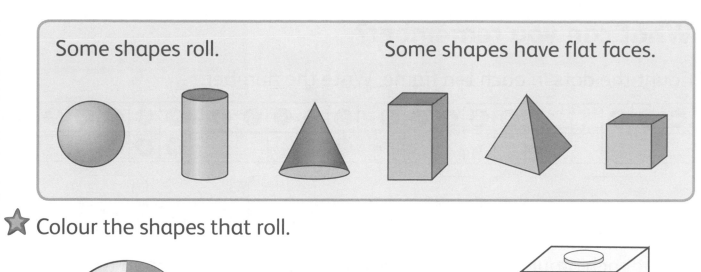

Some shapes roll.　　　　　Some shapes have flat faces.

⭐ Colour the shapes that roll.

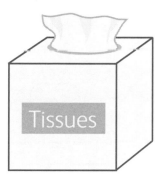

What can you remember?

⭐ Count the dots in each ten frame. Write the number.

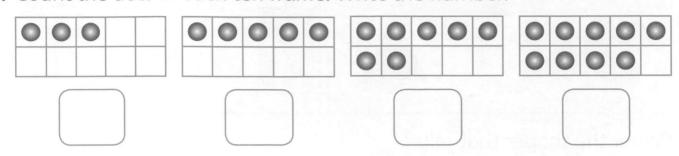

⭐ Count each group of balloons. Draw a line to join each group to its number.

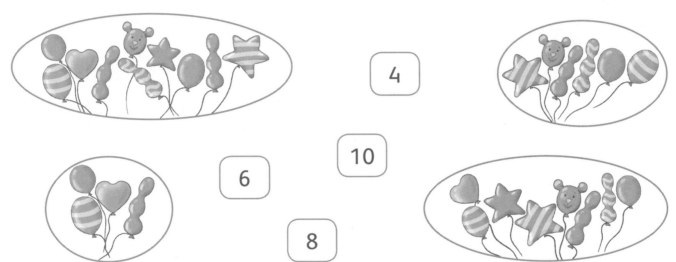

4

10

6

8

⭐ Show the jumps and write the finish number.

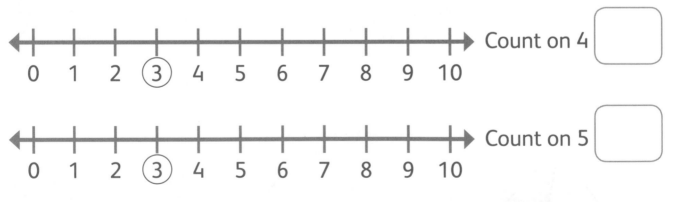

Count on 4

Count on 5

0   1   2   ③   4   5   6   7   8   9   10

0   1   2   ③   4   5   6   7   8   9   10

⭐ Cross out 2 stars in each row. Write how many stars are left.

4 take away 2 is

7 take away 2 is

Number these pictures in order.

 Tick the shortest ribbon. Colour the longest ribbon.

 Draw 3 more shapes in each row.
Colour the shapes to make a pattern.

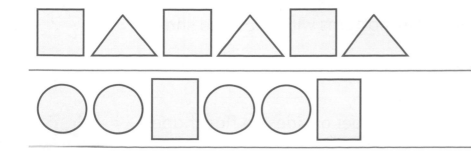

Self-assessment

Colour the stars to show what you can do!

Numbers to 10	I can show numbers to 10 on a ten frame.	☆
	I can write the numbers to 10.	☆
Counting to 10	I can count forwards and backwards on a number track.	☆
	I can work out any missing numbers to 10 on a number track.	☆
	I can count on from different numbers to 10.	☆
Addition and subtraction	I can put two groups of objects together and count the total.	☆
	I can break up 10 cubes in different ways and show the totals.	☆
	I can take away cubes from a set and say what is left.	☆
Time and position	I know the date of my birthday.	☆
	I can describe where an object is.	☆
Measures	I can put objects into order of length from longest to shortest.	☆
	I can put objects into order of size from smallest to largest.	☆
	I can put four jugs in order of capacity, from full to empty.	☆
Shapes and patterns	I can make and draw patterns with lines and shapes.	☆
	I can point to squares, triangles and circles that are around me.	☆
	I can talk about the number of sides on flat shapes.	☆